Ornithology Beginner Guide Made Easy

Understanding the Importance of Studying Birds

By

Adley Brice

Table of Contents

4

CHAPTER 1

Introduction

1.1 What is Ornithology?

Ornithology is the scientific study of birds, encompassing a wide array of disciplines and approaches aimed at understanding the biology, behavior, ecology, evolution, and conservation of these remarkable creatures. Derived from the Greek words "ornis" (bird) and "logos" (study), ornithology is an interdisciplinary field that combines elements of biology, ecology, anatomy, physiology, behavior, genetics, and more to gain insights into the avian world.

Ornithologists, the scientists who specialize in this field, examine the intricate lives of birds, ranging from their physical adaptations and flight mechanics to their breeding behaviors, migration patterns, and interactions with the environment. By delving into the fascinating world of birds, ornithologists contribute to our knowledge of biodiversity, provide insights into the Earth's ecological systems, and contribute to the growing body of evidence on the impacts of climate change and habitat loss.

The study of ornithology is not just limited to scientific research; it extends to birdwatching, an immensely popular recreational activity that has a strong connection to science. Birdwatchers, or birders, are often citizen scientists who contribute valuable data on bird distribution, behavior, and abundance, aiding professional ornithologists in their work.

1.2 History of Ornithology

The history of ornithology dates back to ancient civilizations where birds held symbolic and cultural significance. For example, ancient Egyptians had hieroglyphs depicting birds and often revered certain species like the ibis. In ancient Greece, Aristotle made significant contributions to early ornithological studies with his observations and descriptions of various bird species.

During the Middle Ages, ornithology was often intertwined with folklore and mythology, with limited scientific rigor. However, the Renaissance period saw a rekindled interest in the scientific study of birds. Pioneering figures like John Ray and Francis Willughby laid the foundation for modern ornithology in the 17th century by systematically classifying and describing bird species.

The 18th and 19th centuries witnessed a surge in ornithological exploration, with naturalists and collectors traveling the world to discover and document new species. The development of more advanced technologies, such as binoculars and the microscope, allowed for more detailed observations and descriptions. The advent of photography in the 19th century also revolutionized ornithology by enabling the capture of detailed images of birds in their natural habitats.

In the 20th century, ornithology underwent rapid evolution with the development of advanced techniques in genetics, ecology, and the use of tools like radio telemetry to track bird movements. This era also saw a greater focus on conservation and the understanding of how human activities impact avian populations and their habitats.

Today, ornithology continues to thrive as a dynamic and evolving field, combining traditional field observations with cutting-edge technologies like DNA sequencing, satellite tracking, and remote sensing to answer complex questions about bird biology and ecology.

1.3 Importance of Studying Birds

Studying birds holds immense importance for various reasons:

1. **Biodiversity and Conservation:** Birds represent a diverse group of animals, with over 10,000 species worldwide. By studying them, ornithologists contribute to our understanding of global biodiversity and play a crucial role in conservation efforts. Birds often serve as indicators of ecosystem health, and changes in their populations can signal environmental changes and potential threats to other species, including humans.

2. **Ecological Balance:** Birds play key roles in various ecosystems. They help control insect populations, disperse seeds, and pollinate plants, contributing to ecosystem stability and plant diversity. Studying their interactions with other species and the environment is vital for maintaining healthy ecosystems.

3. **Migration and Climate Change:** Birds undertake incredible migrations, some spanning thousands of miles. The study of bird migration patterns and its connection to climate change provides valuable insights into how environmental factors affect wildlife. Understanding these

patterns can have broader implications for global climate studies.

4. **Behavior and Evolution:** Bird behaviors, such as mating rituals, nest-building, and foraging strategies, are fascinating subjects of study. Ornithologists explore the evolutionary origins of these behaviors, shedding light on the broader field of animal behavior and the evolutionary history of all life on Earth.

5. **Public Engagement:** Birds are easily accessible to the public, making them an ideal subject for citizen science projects. Birdwatching and data collection by enthusiasts contribute to scientific

knowledge and raise awareness about conservation issues.

6. **Inspiration and Education:** Birds have fascinated humans for centuries and continue to inspire art, literature, and scientific discovery. Studying birds is an excellent way to engage young minds in the wonders of nature and foster a love for science and the environment.

ornithology, the scientific study of birds, not only enriches our understanding of these remarkable creatures but also provides invaluable insights into broader ecological and environmental processes, making it a field of study with profound significance for both science and society.

CHAPTER 2

Bird Classification and Taxonomy

2.1 Systematics and Classification

Systematics, within the context of ornithology, is the science of categorizing and organizing the vast diversity of bird species into a logical and informative classification system. This classification system is based on the principles of taxonomy, which is the science of naming, defining, and classifying organisms. The primary goals of systematics and taxonomy in ornithology are to provide a clear and hierarchical framework for understanding the evolutionary relationships among birds and to facilitate the communication of knowledge about different bird species.

Key concepts within the realm of systematics and classification in ornithology include:

- **Phylogenetics:** This is the study of the evolutionary relationships among bird species. Phylogenetic analyses use various forms of data, such as DNA sequences, morphological traits, and behavior, to construct evolutionary trees (phylogenies) that represent the branching patterns of bird lineages.

- **Taxonomic Ranks:** Birds are classified into various

hierarchical ranks, including domain, kingdom, phylum, class, order, family, genus, and species. This hierarchical system allows ornithologists to group birds based on shared characteristics and evolutionary history.

- **Nomenclature:** The naming of bird species follows a binomial nomenclature system, where each species is assigned a unique two-part Latin name, typically consisting of the genus and species names. For example, the common house sparrow is known as Passer domesticus.

- **Type Specimens:** To ensure clarity and consistency, each species is typically associated with a designated type specimen, which serves as a reference point for that species. These type specimens are usually preserved in natural history museums.

- **Checklists and Field Guides:** Ornithologists create checklists and field guides that help birdwatchers and researchers identify and categorize bird species based on their characteristics, distribution, and taxonomy.

2.2 Orders and Families of Birds

Birds are classified into a hierarchical system of orders and families, which represent larger groupings within the avian world. These groupings are based on shared evolutionary history and distinct characteristics. Some common orders and families include:

- **Order Passeriformes (Passerines):** This is the largest order of birds and includes perching birds, such as sparrows, finches, and warblers. Passerines are known for their song and diverse beak shapes, adapted for various diets.

- **Order Falconiformes:** This order includes diurnal birds of prey like hawks, eagles, and falcons, known for their sharp talons and hooked beaks.

- **Order Pelecaniformes:** This order encompasses waterbirds like pelicans, cormorants, and gannets, often found near water bodies.

- **Order Strigiformes:** Owls belong to this order, characterized by their nocturnal habits, large eyes, and silent flight.

- **Family Accipitridae:** This family includes various hawks, eagles, and kites known for their strong beaks and keen eyesight.

- **Family Anatidae:** Waterfowl such as ducks, geese, and swans belong to this family, adapted for swimming and diving in aquatic environments.

- **Family Passeridae:** The sparrow family, known for their small size, brown plumage, and

adaptability to urban
environments.

- **Family Corvidae:** This family
  includes intelligent and highly
  adaptable birds like crows,
  ravens, and magpies.

2.3 Evolution of Birds

The study of bird evolution, often
referred to as avian evolution,
provides insights into the origins and
development of birds as a distinct
group within the animal kingdom.
Some key points about the evolution
of birds include:

- **Dinosaur Ancestry:** Birds are
  believed to have evolved from
  small, bipedal theropod
  dinosaurs during the Mesozoic
  Era, around 150 million years

ago. Archaeopteryx, a fossil from this period, is often considered a transitional form between dinosaurs and modern birds.

- **Feather Evolution:** Feathers, unique to birds, are thought to have initially evolved for insulation and display. Over time, feathers were modified for various functions, including flight, camouflage, and communication.

- **Origin of Flight:** The evolution of powered flight is a defining feature of birds. Modifications in skeletal structure, respiratory systems, and feathers allowed for the development of powered flight.

- **Diversification:** Birds underwent significant diversification, resulting in the myriad species we see today. This diversification was driven by adaptations to various ecological niches, such as aerial predators, aquatic foragers, and terrestrial seed-eaters.

- **Mass Extinction Events:** Throughout their evolutionary history, birds, like many other groups of organisms, were affected by mass extinction events, but they also adapted and radiated into new forms in the aftermath of such events.

Understanding the evolution of birds is critical not only for ornithology but also for shedding light on broader evolutionary processes, such as the development of flight and the relationships between birds and their dinosaur ancestors. This knowledge helps us trace the remarkable journey of birds from their ancient ancestors to the diverse and adaptable group of animals we observe today.

CHAPTER 3

Anatomy and Physiology of Birds

3.1 Skeletal System

The avian skeletal system is adapted for flight, which places specific demands on the structure and function of bones. Key features of the avian skeletal system include:

- **Lightweight Bones:** Birds have relatively lightweight bones with numerous air sacs that reduce overall body weight, making flight more energy-efficient. These air sacs are connected to the respiratory system and help maintain a constant flow of oxygen during both inhalation and exhalation.

- **Fusion and Reduction:** Many bones in birds are fused or reduced in size. For example, the tail is composed of fewer vertebrae than in other vertebrates, and the fingers are often fused to form a single structure that supports the primary flight feathers.

- **Sternum and Keel:** The sternum, or breastbone, features a prominent keel that serves as an anchor for the powerful flight muscles. The large pectoral muscles responsible for wing movement attach to the keel, providing the necessary power for flight.

- **Hollow Bones:** Some of the long bones in birds are hollow, which further reduces weight without compromising strength. These bones have a lattice-like internal structure, enhancing structural integrity.

- **Beak and Skull:** Birds have a beak or bill, adapted for various feeding strategies, such as pecking, probing, or tearing. The skull is relatively lightweight, with adaptations for jaw movement and beak strength depending on the diet of the bird.

3.2 Muscular System

Birds possess highly developed muscular systems, particularly in relation to flight and locomotion.

Some key aspects of the avian muscular system include:

- **Flight Muscles:** Birds have two major sets of flight muscles, the pectoral muscles and the supracoracoideus muscles. The pectoral muscles, attached to the keel of the sternum, provide the downstroke during flight, while the supracoracoideus muscles control the upstroke.

- **Cardiac Muscle:** The heart of birds, like other animals, is primarily composed of cardiac muscle, which powers the circulatory system. Birds have high metabolic rates, and their hearts are adapted to efficiently deliver oxygen to muscles during flight.

- **Leg Muscles:** The legs of birds are well-muscled, with adaptations for various forms of locomotion. Birds can have powerful leg muscles for walking and running, as seen in ostriches and emus, or specialized talons for grasping, as in raptors.

- **Neck Muscles:** Birds have a flexible neck with strong muscles that allow for precise movements and manipulation of the head and beak. This is essential for tasks like grooming, feeding, and preening.

- **Sound Production:** Some birds, such as songbirds, have specialized syringeal muscles that control the syrinx, the avian vocal organ. These muscles play a role in sound production and song.

3.3 Respiratory System

The avian respiratory system is unique and highly efficient, enabling birds to meet the high oxygen demands of flight. Key features of the avian respiratory system include:

- **Air Sacs:** Birds have a system of air sacs in addition to their lungs. These air sacs act as bellows, constantly moving air through the respiratory system, even during both inhalation and exhalation. This continuous flow of air maintains a high oxygen supply.

- **Posterior and Anterior Air Sacs:** Birds have both posterior and anterior air sacs. The posterior air sacs are located near the posterior end of the bird, while the anterior air sacs are close to the front. Air flows through the lungs twice during a complete cycle of inhalation and exhalation, which enhances gas exchange efficiency.

- **Parabronchi:** The avian lung is made up of a series of interconnected air sacs and tubes called parabronchi, where gas exchange occurs. This arrangement allows for a unidirectional flow of air through the lung, optimizing oxygen uptake.

- **High Metabolic Rate:** The avian respiratory system is adapted to support the high metabolic rate required for flight. Birds have a significantly higher oxygen demand compared to many other animals of similar size.

- **Supplemental Air Sacs:** In some species, supplemental air sacs extend into various parts of the body, such as the bones. These sacs may play a role in temperature regulation, vocalization, or buoyancy control.

The unique adaptations in the skeletal, muscular, and respiratory systems of birds reflect their specialization for flight and diverse ecological roles, making them one of the most successful groups of animals on Earth. These adaptations have allowed birds to colonize a wide range of habitats and exhibit remarkable diversity in form and function.

3.4 Circulatory System

The avian circulatory system plays a crucial role in supplying oxygen and nutrients to the body's cells, supporting the high metabolic demands of birds, particularly during flight. Key features of the avian circulatory system include:

- **Four-Chambered Heart:** Birds have a four-chambered heart, similar to mammals, consisting of two atria and two ventricles. This separation of oxygenated and deoxygenated blood allows for efficient oxygen delivery to the body tissues.

- **High Heart Rate:** Birds typically have rapid heart rates, which can vary depending on the species. The heart rate increases during flight to meet the increased oxygen demand.

- **Systemic and Pulmonary Circulation:** Blood is pumped from the heart to both systemic (body) and pulmonary (lung) circulations. The unique arrangement of air sacs and parabronchi in the respiratory system enhances gas exchange and oxygen delivery.

- **Efficient Oxygen Transport:** Hemoglobin in avian blood has a higher oxygen-carrying capacity than that of many mammals, which is essential for oxygen delivery during high-energy activities like flying.

3.5 Digestive System

The avian digestive system is adapted to meet the specific dietary needs of different bird species. Key features of the avian digestive system include:

- **Crop:** Birds have a crop, a pouch-like structure in the esophagus, which allows for temporary food storage. This is especially useful for species that need to consume large quantities of food in a short time.

- **Stomach:** Birds have a two-part stomach: the proventriculus, which secretes digestive enzymes, and the gizzard, which grinds food with the help of small, ingested stones (gastroliths).

- **Cecum:** Some birds possess a cecum, a pouch-like structure in the digestive system, where fermentation of plant material occurs. This helps break down complex carbohydrates.

- **Cloaca:** The cloaca is a common chamber where the digestive, excretory, and reproductive systems meet. It serves as a single exit point for waste elimination and reproduction.

- **Beak Adaptations:** The shape and structure of a bird's beak are adapted to its specific diet. Beaks can be adapted for capturing insects, tearing meat, or crushing seeds.

3.6 Nervous System

The avian nervous system is highly developed and well-suited for the complex behaviors and sensory processing required for survival and flight. Key features of the avian nervous system include:

- **Large Brain:** Birds have relatively large brains compared to body size, particularly in regions associated with problem-solving, learning, and complex behaviors. This is evident in species like parrots and crows.

- **Well-Developed Senses:** Birds have well-developed senses, including excellent vision, acute hearing, and a keen sense of smell (in some species). The optic lobes of the brain are particularly large, reflecting the importance of visual information in navigation and foraging.

- **Cerebellum:** The cerebellum is crucial for coordination of muscle movements, which is essential for precise flight control.

- **Learning and Memory:** Many birds exhibit advanced learning and memory abilities, which are vital for tasks such as navigation, mate selection, and foraging. Some species, like parrots and songbirds, are known for their complex vocal learning.

- **Central Pattern Generators:** Birds have central pattern gencrators in their spinal cords

that control rhythmic behaviors,
like walking and flying,
without requiring constant
input from the brain.

3.7 Reproductive System

The avian reproductive system is
highly specialized, and it varies across
species, reflecting different breeding
strategies. Key features of the avian
reproductive system include:

- **Internal Fertilization:** Birds
  have internal fertilization,
  where sperm from the male
  fertilizes eggs within the
  female's reproductive tract.

- **Oviparous:** Birds are
  oviparous, meaning they lay
  eggs rather than giving birth to
  live young. The structure and

number of eggs vary among species.

- **Eggshell Formation:** The avian reproductive system includes the development of a calcified eggshell, which protects the developing embryo. The shell's thickness and texture vary based on the species and nesting habits.

- **Brood Patches:** Some female birds develop specialized patches of bare skin on their abdomens, called brood patches, which help transfer heat from the parent's body to the eggs during incubation.

- **Parental Care:** The level of parental care varies widely among bird species. Some species exhibit extensive care, with both parents feeding and protecting the young, while others employ different strategies, such as communal nesting or parasitism.

The avian reproductive system reflects the remarkable diversity of bird species and their adaptations to various environments and lifestyles, from ground-nesting birds to those that build intricate nests in trees or cliffs. These adaptations ensure the survival of offspring in a range of ecological niches.

CHAPTER 4

Bird Behavior

4.1 Feeding Strategies

Birds exhibit a wide range of feeding strategies, which are closely tied to their beak and bill morphology, as well as their ecological niches. Some common feeding strategies include:

- **Carnivory:** Carnivorous birds, like raptors (eagles, hawks), owls, and many songbirds, hunt and eat other animals, such as insects, small mammals, fish, and other birds. Their sharp beaks and talons are adapted for capturing and killing prey.

- **Herbivory:** Herbivorous birds, such as parrots and doves, primarily consume plant material like seeds, fruits, nectar, and leaves. Their beaks are adapted for cracking seeds, extracting nectar, or tearing plant matter.

- **Omnivory:** Omnivorous birds have a varied diet that includes both plant and animal matter. For example, crows and gulls scavenge for carrion, eat small invertebrates, and consume human food.

- **Filter-Feeding:** Some birds, like flamingos and some ducks, have specialized bills for filter-feeding. They feed by straining small organisms, algae, and plankton from the water.

- **Insectivory:** Insectivorous birds, including flycatchers and warblers, primarily feed on insects and other arthropods. They often catch insects in mid-air or glean them from foliage.

- **Scavenging:** Scavengers, such as vultures, play an essential ecological role by consuming carrion. Their strong immune systems allow them to consume decaying meat.

4.2 Migration

Migration is a remarkable behavior seen in many bird species. It involves regular, seasonal movements between breeding and non-breeding areas. Key aspects of bird migration include:

- **Long-Distance Travel:** Many migratory birds cover thousands of miles during migration, sometimes crossing continents or even hemispheres. This behavior is often driven by the need to find suitable breeding and wintering grounds.

- **Navigation:** Birds use various cues for navigation, including visual landmarks, the Earth's magnetic field, the position of the sun and stars, and even their innate genetic programming. Some species, like homing pigeons, have exceptional navigation abilities.

- **Energetic Demands:** Migration requires significant energy, and birds often need to replenish their energy reserves by feeding along the way. Stopover sites are critical for rest and refueling during migration.

- **Timing:** Migration is typically timed with seasonal changes in temperature, food availability, and day length. Birds migrate south in the fall as temperatures drop and resources become scarcer, and they return north in the spring as conditions become more favorable for breeding.

4.3 Communication and Vocalization

Birds use a variety of vocalizations and other forms of communication for various purposes:

- **Song:** Songbirds (passerines) are known for their complex songs, which serve multiple functions, including territory defense, mate attraction, and communication with other members of the same species.

- **Calls:** Birds use calls for a wide range of purposes, such as warning of predators, maintaining contact with a flock, and coordinating activities within a group.

- **Visual Signals:** Some birds use visual displays to communicate, including courtship rituals, feather displays, and body postures.

- **Mimicry:** Certain species, like the superb lyrebird, are renowned for their ability to mimic other bird species and sounds in their environment.

- **Duets:** Some bird species engage in duets, where both males and females contribute to coordinated vocalizations. This behavior often strengthens pair bonds.

- **Communication in Social Groups:** In species that exhibit complex social behaviors, communication is vital for group cohesion, foraging efficiency, and predator avoidance.

4.4 Breeding Behavior

Breeding behavior in birds varies greatly among species, and it is influenced by ecological factors, reproductive strategies, and the availability of resources. Some common aspects of breeding behavior include:

- **Courtship Displays:** Many bird species engage in elaborate courtship displays to attract mates. These displays may involve singing, dancing, or presenting gifts.

- **Nest Building:** Birds construct nests to protect their eggs and young. Nest construction varies widely, from simple scrapes on the ground to intricate structures in trees or cliffs.

- **Egg Laying:** The timing of egg laying varies among species. Some birds lay a single egg at a time, while others lay clutches of multiple eggs. The size, shape, and color of eggs can also differ significantly.

- **Incubation:** Incubation of eggs is typically the responsibility of one or both parents. The incubation period varies depending on the species.

- **Parental Care:** Parental care can involve feeding, protecting, and teaching offspring. Some birds provide extensive care, while others are precocial and can feed themselves shortly after hatching.

- **Brood Parasitism:** In some cases, birds employ brood parasitism, where they lay their eggs in the nests of other bird species, shifting the responsibility of rearing their young to the host species.

4.5 Social Behavior

Birds exhibit a range of social behaviors, from solitary to highly social. Key aspects of social behavior in birds include:

- **Solitary Species:** Some birds, like owls and eagles, are primarily solitary, coming together only for breeding and territorial defense.

- **Colonial Nesting:** Many seabirds, such as gulls and terns, nest in large colonies for protection and to exploit concentrated food resources.

- **Flocking:** Flocking behavior is common in various species, including starlings and swallows. Flocking provides benefits such as safety from predators and increased foraging efficiency.

- **Territoriality:** Many bird species are territorial, defending specific areas for breeding, foraging, or roosting. Territorial disputes often involve vocalizations and displays.

- **Cooperative Breeding:** In some species, individuals other than the breeding pair help raise young. These helpers might be offspring from previous broods or non-breeding individuals.

- **Mating Systems:** Mating systems can vary, from monogamy (pair bonding) to polygamy (having multiple mates). The choice of mating system often relates to

ecological and resource
availability.

Bird behavior is a fascinating and
diverse subject of study, offering
insights into the evolution of different
species, their adaptations to ecological
niches, and the complex interactions
that shape their lives. Understanding
these behaviors is vital for bird
conservation and the preservation of
biodiversity.

CHAPTER 5

Bird Ecology

5.1 Habitats and Ecosystems

Birds inhabit a wide range of habitats and ecosystems, each with its own unique set of environmental conditions and resources. Some key points regarding bird ecology in various habitats and ecosystems include:

- **Forest Ecosystems:** Many bird species are found in forests, where they have adapted to life in dense vegetation. Different types of forests, such as temperate, tropical, and coniferous forests, host distinct bird communities. Birds in forests play roles in seed dispersal, insect control, and shaping the structure of the forest.

- **Grassland Ecosystems:** Grasslands are home to numerous bird species, from sparrows to raptors. Birds in grasslands are often ground-nesters and forage for insects, seeds, or small mammals. Some grassland birds are highly migratory, traveling long distances between breeding and wintering grounds.

- **Wetland Ecosystems:** Wetlands, including marshes, swamps, and estuaries, support a diverse bird community. Waterbirds, like herons and ducks, are common in wetlands, where they feed on fish, invertebrates, and aquatic plants. Wetlands also serve as important stopover points for migratory birds.

- **Desert Ecosystems:** Birds in desert ecosystems have evolved various adaptations to cope with extreme heat and aridity. These adaptations include the ability to go without water for extended periods, specialized bills for feeding on seeds and insects, and efficient heat dissipation mechanisms.

- **Urban and Suburban Ecosystems:** Some bird species have adapted to human-altered environments. Urban and suburban areas may support a range of birds, from pigeons and sparrows to raptors like peregrine falcons. These birds utilize buildings, parks, and gardens for nesting and foraging.

- **Mountain Ecosystems:** High-altitude mountain ecosystems are inhabited by specialized birds adapted to low oxygen levels and cold temperatures. These include species like snow finches and snow leopards. Mountain birds are often characterized by their vibrant plumage and unique behaviors.

- **Island Ecosystems:** Islands, whether oceanic or freshwater, often support unique bird species found nowhere else. Island birds may exhibit flightlessness, large sizes, or distinct plumage patterns due to the lack of terrestrial predators.

- **Mangrove Ecosystems:** Mangroves are important for various bird species, including wading birds and shorebirds. These areas provide shelter and abundant food resources in the form of fish and invertebrates.

- **Cave Ecosystems:** Certain bird species, like swifts and swallows, roost in caves or cave-like structures. These birds are adapted for aerial foraging and are often associated with cave ecosystems due to the availability of suitable roosting sites.

5.2 Feeding Ecology

Birds have diverse feeding strategies that are closely linked to their beak and bill structures, as well as the availability of food resources in their specific habitats. Feeding ecology includes various aspects of bird diets and foraging behaviors:

- **Carnivores:** Carnivorous birds primarily feed on animal matter, which can include insects, fish, small mammals, and other birds. Raptors, such as eagles and hawks, are well-known carnivorous birds.

- **Herbivores:** Herbivorous birds feed on plant material, including seeds, fruits, nectar, and leaves. This group includes species like parrots, doves, and finches.

- **Omnivores:** Omnivorous birds have diets that encompass both plant and animal matter.

Examples include crows, gulls, and some songbirds that feed on a variety of food sources.

- **Insectivores:** Insectivorous birds specialize in feeding on insects and other invertebrates. They may capture insects in flight, glean them from vegetation, or search for them under bark or in the soil.

- **Frugivores:** Frugivorous birds are fruit eaters, often playing a role in seed dispersal. Birds like toucans and hornbills have specialized bills for consuming fruit.

- **Nectarivores:** Nectarivorous birds feed on nectar, often from flowers. They have specialized long, slender bills and extendable, brush-tipped tongues for extracting nectar. Examples include hummingbirds and sunbirds.

- **Filter Feeders:** Some waterbirds, like flamingos and certain species of ducks, are filter feeders. They strain tiny aquatic organisms and algae from the water using specialized structures in their bills.

- **Scavengers:** Scavengers, such as vultures and crows, feed on carrion, helping to recycle nutrients in ecosystems.

- **Specialized Foragers:** Birds have evolved diverse foraging behaviors and feeding adaptations, such as tool use, cooperative hunting, and trap-building. These strategies enable them to exploit specific niches and food sources in their environments.

Birds are integral to the balance of ecosystems as they shape plant and animal populations through predation, pollination, seed dispersal, and other ecological roles. Understanding their feeding ecology is crucial for conserving both bird species and their associated ecosystems.

5.3 Nesting and Reproductive Ecology

Nesting and reproductive ecology in birds are diverse and highly adapted to the specific ecological niches in which different species live. Key aspects of nesting and reproductive ecology include:

- **Nest Types:** Birds construct a wide variety of nests, from simple scrapes on the ground to intricate structures in trees, cliffs, burrows, or even on man-made structures. Nests provide shelter and protection for eggs and chicks.

- **Egg Laying:** The timing of egg laying varies among species and is often influenced by factors like temperature, food availability, and day length. Some birds lay one egg at a time, while others lay clutches of multiple eggs.

- **Incubation:** Incubation of eggs is typically the responsibility of one or both parents, with variations depending on the species. Incubation duration varies based on environmental conditions and the size of the bird.

- **Parental Care:** Parental care encompasses feeding, protecting, and teaching offspring. Parental care behaviors vary widely, with some species providing extensive care, such as penguins and eagles, while others are precocial and can feed themselves shortly after hatching.

- **Brood Parasitism:** In some cases, birds employ brood parasitism, where they lay their eggs in the nests of other bird species, shifting the responsibility of rearing their young to the host species.

- **Mate Selection:** Courtship rituals and displays are crucial for mate selection. These behaviors can involve singing, dancing, or presenting gifts to attract potential mates.

- **Nesting Success:** Nesting success, defined as the proportion of nests that successfully fledge offspring, is influenced by various factors, including predation, food availability, and habitat quality.

- **Seasonal Breeding:** Many bird species breed seasonally to coincide with the availability of food resources for their offspring. This timing can vary with latitude and climate.

5.4 Conservation and Threats to Bird Habitats

Birds face numerous threats to their habitats and survival, and conservation efforts are essential to protect bird populations and the ecosystems they inhabit. Common threats to bird habitats include:

- **Habitat Loss:** Habitat destruction through deforestation, urbanization, agriculture, and land development is a major threat to bird species. Many birds require specific types of habitats for breeding, foraging, and roosting.

- **Climate Change:** Climate change is affecting the distribution and behavior of bird species. Birds may face challenges in adapting to shifting climate patterns, including altered migration timing and mismatches with food availability.

- **Pollution:** Pollution, including pesticide use, oil spills, and contamination of water sources, can have harmful effects on bird populations. Some chemicals accumulate in the food chain and can lead to reproductive problems and declines in bird species.

- **Invasive Species:** Invasive plants and animals can disrupt native ecosystems and outcompete native bird species for resources. For example, invasive predators can impact nesting success.

- **Hunting and Poaching:**
  Hunting and poaching can lead
  to population declines in some
  bird species. Legal and illegal
  hunting for food, sport, and
  trade poses a significant threat
  to birds in certain regions.

- **Overexploitation:**
  Overharvesting of bird species
  for their feathers, eggs, or pets
  has led to declines in some
  populations. This has led to the
  implementation of international
  agreements to protect these
  species.

- **Nesting Disturbance:** Human
  disturbance, such as
  approaching nesting sites, can
  lead to nest abandonment and
  reduced nesting success.
  Conservation efforts often

focus on minimizing such disturbances.

- **Disease:** Disease outbreaks, often facilitated by human activity and global travel, can affect bird populations. Avian influenza and West Nile virus are examples of diseases that have had an impact on bird species.

Conservation efforts to mitigate these threats include habitat protection, restoration, and management, captive breeding and reintroduction programs, efforts to control invasive species, and public education about the importance of bird conservation. International agreements, such as the Migratory Bird Treaty Act and the Convention on International Trade in Endangered Species (CITES), aim to safeguard bird populations and their habitats.

Birdwatchers and citizen scientists also play a vital role in monitoring and conserving bird populations by providing valuable data for research and conservation initiatives.

CHAPTER 6

Bird Conservation

6.1 Threats to Bird Populations

Bird populations face a range of threats that can have significant impacts on their numbers and distribution. Some of the key threats to bird populations include:

- **Habitat Loss and Degradation:** The conversion of natural habitats into urban areas, agriculture, and infrastructure development results in habitat loss and degradation. This threatens bird species that rely on specific habitats for breeding, foraging, and roosting.

- **Climate Change:** Global climate change is altering the distribution and behavior of bird species. Rising temperatures, shifts in seasonal timing, and changes in precipitation patterns can affect food availability and breeding success.

- **Pollution:** Pollution from various sources, including agricultural runoff, industrial discharge, and contaminants in water bodies, can negatively impact bird populations. Pesticides and chemicals can harm birds directly or indirectly through the food chain.

- **Invasive Species:** Invasive plants and animals can outcompete native species for resources and disrupt ecosystems. Invasive predators, like rats and feral cats, pose a particular threat to nesting birds and their young.

- **Overexploitation:** Overharvesting of birds for food, pets, or their body parts can lead to population declines. Some species are hunted or trapped in unsustainable ways.

- **Nesting Disturbance:** Human disturbance, such as approaching nesting sites or recreational activities near sensitive areas, can cause nest abandonment and decreased nesting success.

- **Collisions:** Birds can collide with buildings, vehicles, and power lines, resulting in injuries or fatalities. These collisions are a significant issue in urban and developed areas.

- **Disease:** Disease outbreaks, often facilitated by human activity and global travel, can affect bird populations. Avian influenza, West Nile virus, and avian malaria are examples of diseases that impact bird species.

- **Light Pollution:** Artificial light at night can disorient migrating birds, leading to collisions with buildings and other structures. It can also affect nocturnal behaviors, including hunting and mating.

- **Illegal Trade:** The illegal trade in birds, their eggs, and their body parts remains a concern. Some bird species are highly sought after in the pet trade, despite international regulations.

6.2 Conservation Efforts

Efforts to conserve bird populations and their habitats are multifaceted and often involve government agencies, non-governmental organizations, and concerned individuals. Conservation initiatives include:

- **Habitat Protection:** Establishing and maintaining protected areas, such as national parks and wildlife reserves, is critical for conserving essential bird habitats. These areas can also serve as refuges for threatened species.

- **Habitat Restoration:** Restoration projects aim to restore degraded habitats to their natural state. This can include reforestation, wetland restoration, and removal of invasive species.

- **Conservation Breeding:** Captive breeding and reintroduction programs are used to bolster the populations of endangered or threatened bird species. These programs may be carried out in cooperation with zoos and aviaries.

- **Invasive Species Control:** Managing and controlling

invasive species is essential for the protection of native birds. This may involve the removal of invasive plants and the establishment of predator control programs.

- **Conservation Agreements:** International agreements, such as the Migratory Bird Treaty Act, the Convention on International Trade in Endangered Species (CITES), and the Convention on the Conservation of Migratory Species of Wild Animals (CMS), help regulate and protect bird species across borders.

- **Public Awareness and Education:** Increasing public awareness of the importance of bird conservation is vital.

Conservation organizations and educational programs work to engage the public in bird conservation efforts.

- **Research and Monitoring:** Ongoing research and monitoring of bird populations and their habitats provide valuable data for conservation decision-making. Citizen science initiatives play a crucial role in data collection.

- **Advocacy and Policy:** Advocacy efforts seek to influence government policies and regulations to better protect birds and their habitats. This may include advocating for stronger environmental laws and regulations.

- **Collaboration:** Collaborative efforts between governments, conservation organizations, researchers, and local communities are often crucial for the success of conservation initiatives.

6.3 Citizen Science and Birdwatching

Citizen science and birdwatching play a vital role in bird conservation by involving the public in data collection, monitoring, and advocacy. These activities have several benefits:

- **Data Collection:** Birdwatchers and citizen scientists contribute valuable data on bird populations, distributions, and behaviors. This information helps researchers and conservationists make informed decisions.

- **Biodiversity Monitoring:** By monitoring bird species and populations, citizens can assess the health of ecosystems and the impact of conservation efforts.

- **Engagement and Education:** Birdwatching and citizen science provide opportunities for people to connect with nature and develop an appreciation for birds and their conservation. This engagement can lead to increased support for conservation initiatives.

- **Advocacy:** Informed citizens can become advocates for bird conservation and environmental protection, helping to influence policies and raise awareness about the importance of preserving bird habitats.

- **Community Involvement:** Citizen science and birdwatching often foster a sense of community and cooperation among individuals interested in bird conservation, creating a network of like-minded people working toward a common goal.

Popular citizen science initiatives that involve birdwatching include the Audubon Christmas Bird Count, the Great Backyard Bird Count, and eBird, which encourage bird enthusiasts to record their observations and contribute to the scientific understanding of bird populations and migration patterns.

CHAPTER 7

Field Methods in Ornithology

7.1 Bird Identification

Bird identification is a fundamental skill in ornithology, allowing researchers and enthusiasts to distinguish between bird species. Several methods and tools are used for bird identification:

- **Field Guides:** Field guides are books that provide illustrations, photographs, and detailed descriptions of bird species. They help users identify birds based on their physical characteristics, including size, plumage coloration, bill shape, and other distinguishing features.

- **Binoculars:** Binoculars are essential for birdwatching and ornithological research, as they allow observers to get a closer view of birds in the field. Optics quality, such as magnification and lens coatings, is important for accurate identification.

- **Spotting Scopes:** Spotting scopes are high-powered telescopes that provide even more magnification than binoculars. They are especially useful for observing distant or small birds.

- **Bird Calls and Songs:** Bird vocalizations are crucial for identifying species. Many birdwatchers and ornithologists learn to recognize the songs and calls of various birds, as these can be just as distinctive as visual features.

- **Smartphone Apps:** Several smartphone apps are available for bird identification, often using image recognition or audio matching to help users identify birds in the field.

- **Museum Collections:** Museums and research institutions maintain collections of bird specimens that can be used for comparison and identification. These collections provide a reference for bird morphology and plumage.

7.2 Bird Banding and Mark-Recapture

Bird banding, also known as bird ringing, is a technique used to track individual birds and gather data on their movements, lifespan, and population dynamics. Here's how it works:

- **Banding:** A small, individually numbered metal or plastic band is placed on a bird's leg. These bands are provided by organizations like the U.S. Geological Survey (USGS) in the United States.

- **Mark-Recapture:** Researchers capture birds, record the band number, and collect data on the bird's age, sex, weight, and other characteristics. The bird is then released. If the same bird is recaptured at a later time, the information provides insights into migration, longevity, and population size.

- **Population Estimation:** Mark-recapture data can be used to estimate population size and demographics. Different statistical models are applied to account for variations in recapture rates and capture probabilities.

- **Migration Studies:** Bird banding is instrumental in tracking the migrations of many species. It helps researchers understand migration routes, stopover locations, and the timing of migrations.

- **Longevity and Life History:** Banding data can reveal

information about the age, reproductive success, and survival rates of individual birds.

- **Research and Conservation:** Bird banding is a valuable tool for studying bird ecology, behavior, and conservation. It provides critical information for the management of endangered or threatened species.

7.3 Survey and Census Techniques

Survey and census techniques are used to assess bird populations and gather data on species abundance and distribution. Several methods are employed in ornithological research:

- **Point Counts:** Researchers or birdwatchers visit specific locations and count all the birds they see or hear within a set time period (e.g., 5 minutes). Point counts are used to monitor bird populations over time.

- **Transect Surveys:** Transect surveys involve walking along predetermined paths (transects) and recording all bird species seen or heard within a certain distance on either side of the path. This method is often used for assessing habitat use and species richness.

- **Nocturnal Surveys:** Nighttime surveys, often conducted using auditory techniques like listening for owl calls, are used to study nocturnal birds and owls. Bat detectors are sometimes employed for recording echolocation calls of insectivorous bats.

- **Pellet Analysis:** The analysis of regurgitated pellets produced by birds of prey, such as owls, can provide valuable insights into their diet and prey selection.

- **Nest Monitoring:** Researchers monitor bird nests to study nesting success, reproductive behavior, and chick survival. Nest boxes are sometimes used to facilitate monitoring.

- **Camera Traps:** Camera traps equipped with motion sensors are used to capture images or videos of birds and their behavior. They are particularly useful for elusive or shy species.

- **Radar and Acoustic Technology:** Remote sensing technologies, including radar and acoustic monitoring, are employed to study bird movements, migrations, and behavior on a larger scale.

- **Bioacoustic Monitoring:** Bioacoustic recorders capture bird vocalizations and songs. This technology allows for the remote monitoring of bird presence and behavior.

- **Geospatial Techniques:** Geographic Information Systems (GIS) and satellite telemetry are used to track bird movements, migration routes, and habitat use on a large scale.

The choice of survey and census technique depends on the research goals, the species of interest, and the habitat being studied. A combination of methods is often used to gather a comprehensive understanding of bird populations and their ecological roles.

CHAPTER 8

Ornithology in the Modern World

8.1 Avian Research and Technology

Modern ornithology benefits from a wide range of advanced technologies that have transformed the field. These technologies enhance our ability to study birds, monitor populations, and gather data more effectively. Some notable advancements include:

- **Avian Tracking Devices:** Small, lightweight GPS trackers and satellite transmitters allow researchers to track bird movements with unprecedented precision. These devices provide insights into migration routes, stopover locations, and wintering grounds.

- **Bioacoustics:** Digital audio recorders and software for bioacoustic analysis have made it easier to record and analyze bird vocalizations. This technology aids in the study of bird communication, behavior, and population monitoring.

- **Remote Sensing:** Remote sensing technologies, including satellite imagery and LiDAR (Light Detection and Ranging), provide valuable data for habitat analysis and landscape-scale research. They are especially useful for understanding bird habitat use and distribution.

- **Genomic Tools:** Advances in genomics have facilitated genetic studies in ornithology. Techniques like DNA sequencing and genotyping have contributed to our understanding of avian genetics, evolution, and population structure.

- **Camera Traps:** Motion-activated cameras, often used for studying terrestrial wildlife, are increasingly applied to ornithological research. They capture images and videos of birds and their behavior in various settings.

- **Drones:** Unmanned aerial vehicles (drones) equipped with cameras and sensors offer a non-invasive means of studying birds and their nests. They are used for monitoring and habitat assessment.

- **Geospatial Analysis:** Geographic Information Systems (GIS) and spatial analysis software are essential for modeling bird distributions, assessing habitat changes, and conducting landscape-level studies.

- **E-DNA Sampling:** Environmental DNA (e-DNA) analysis allows for the detection of bird species through the genetic material they leave in the environment, such as feathers, feces, or saliva. This non-invasive technique is useful for studying elusive or rare species.

8.2 Current Trends in Ornithology

Ornithology is a dynamic field that reflects current ecological and conservation concerns. Some prominent trends in modern ornithology include:

- **Climate Change and Phenology:** Ornithologists are increasingly studying how climate change affects bird phenology—seasonal events such as migration, breeding, and foraging. Shifts in timing and range expansions are observed.

- **Urban Ornithology:** With growing urbanization, urban ornithology has become a focus. Researchers study how birds adapt to urban environments and how urbanization impacts their populations.

- **Citizen Science:** Citizen science projects, such as eBird and Christmas Bird Counts, have grown in popularity, involving the public in data collection and conservation efforts. This crowdsourced data is invaluable for monitoring bird populations.

- **Conservation Technology:** Conservation efforts often involve technology, from captive breeding programs to the use of drones for habitat monitoring and anti-poaching activities in remote areas.

- **Invasive Species Management:** Ornithologists are involved in efforts to control and manage invasive species that threaten native bird populations. This includes the removal of invasive predators and the restoration of native habitats.

- **Community-Based Conservation:** Collaboration with local communities is increasingly emphasized in bird conservation. Community-based conservation projects often address the needs and concerns of local people while protecting bird habitats.

- **Restoration Ecology:**
  Ornithologists are involved in
  habitat restoration efforts, such
  as wetland restoration,
  reforestation, and ecosystem
  rehabilitation, to benefit both
  birds and ecosystems.

- **Technological Integration:**
  The integration of different
  technologies, such as
  combining tracking data with
  bioacoustic recordings and
  remote sensing imagery, allows
  for more comprehensive and
  interdisciplinary research.

8.3 Future Directions in Bird Research

The future of ornithology holds
exciting opportunities and challenges:

- **Molecular Advances:** Continued advancements in molecular biology will enable in-depth studies of avian genetics, evolution, and the impacts of genetic diversity on population health.

- **Climate Change and Conservation:** Ornithologists will play a key role in understanding and mitigating the impacts of climate change on bird populations, such as altered migration patterns, habitat shifts, and phenological changes.

- **AI and Machine Learning:** Artificial intelligence and machine learning will play a growing role in the analysis of large datasets, such as automated bird recognition in bioacoustic recordings and the processing of remotely sensed imagery.

- **Global Collaboration:** International collaboration and data sharing will be crucial for addressing global conservation challenges, such as migratory bird conservation and the protection of shared habitats.

- **Community Engagement:** Ornithology will increasingly involve local communities in conservation and research efforts, recognizing the importance of indigenous knowledge and community-based initiatives.

- **Emerging Diseases:** The study of emerging diseases affecting birds, such as avian influenza and West Nile virus, will remain a priority, as will understanding the impact of diseases on bird populations.

- **Habitat Restoration and Ecosystem Health:** Ornithologists will continue to contribute to habitat restoration efforts and the broader understanding of ecosystem health, recognizing the interconnectedness of all species.

The future of ornithology will be shaped by a combination of innovative technology, interdisciplinary collaboration, and the collective efforts of ornithologists, birdwatchers, and the public in preserving the rich diversity of bird life on our planet.